Sky's Time

Author: Lisa Durley

Illustrator: Phuong Tran

Editor: Vanessa Durley

Acknowledgements

A Special thanks goes to God for giving me the talent to make me creative and overflowing my life with love, joy and happiness. I would also like to thank my mom, (Glinda Durley), solely for being you and instilling pure laughter into my life. Lastly, my friends and family. No matter what I try to do, you all are always right there to take this Journey with me. I hope everyone enjoys this book!

Sky!

"Yes mom!?" Sky replied, while brushing her teeth.

"You have two hours before Jaylen's birthday party starts." her mom announced

"Okay mom. I will be ready" Sky responded. She glanced at the clock.

A puzzled look crossed Sky's face. She realized that she does not know what time it would be two hours from now.

As she stared out her window, she saw Anthony the ant. 'Maybe I should go ask Anthony the ant' she thought to herself.

 She grabbed a pencil and notepad then glided down the stairs. She raced out the front door to catch her friend.

"Anthony, Anthony" Sky yelled! "I need your help!"

"Anything for you" Anthony replied.

"Mom said we are leaving for Jaylen's party in two hours."

"Okay?" Anthony stated.

"I do not know what two hours from now is" Sky confessed.

"All I can tell you is that there are 60 seconds in 1 minute" Anthony responded.

"60 seconds?" Sky questioned.

"Yes, 60 seconds" Anthony confirmed.

"I hope that helps," Anthony added with a smile then he ran off into the grass.

Sky pulled her pencil and notepad out and wrote:

60 seconds = 1 minute

Sky looked across the street to see Trenton the tiger.

"Trenton, Trenton!" Sky yelled "I need your help!"

A big yawn crossed Trenton's face as he asked, "How may I help you, Sky?"

"I am trying to find out what two hours is, but all I have is

60 seconds= 1 minute" Sky said.

"I can help you with that!" Trenton said excitingly. "There are 60 minutes = 1 hour."

Sky pulled out her notepad and wrote:

60 minutes =1 hour

Trenton slipped off into the woods.

60 seconds = 1 minute
60 minutes = 1 hour

You write it

Sky's excitement quickly faded away when she noticed Trenton was nowhere to be found. Sky slowly slumped down at the foot of her tree trunk. As tears slowly rolled down her face.

Journey the Jaguar noticed the tears on Sky's face and went to her friend. Journey gently tapped Sky on her shoulder and Sky jumped.

"Oh Journey!" Sky smiled as she greeted Journey with a hug.

" I am trying to learn how long is two hours and I all have is:

60 seconds = 1 minute and

60 minutes = 1 hour."

"Smile Sky," Journey said with cheer.

"If

60 seconds = 1 minute and

60 minutes = 1 hour

Then all we have to do is double the minutes."

"Double the minutes?" Sky questioned.

"Yes" Journey explained.
"60 seconds = 1 minute and
60 minutes = 1 hour
120 minutes = 2 hours.

60 seconds = 1 minute
60 minutes = 1 hour
120 minutes = 2 hours

"Oh Journey!" Sky screamed.

"Thank you, thank you, THANK YOU!!!"

Sky danced, flipped and sung all the way up the tree.

"Mom, Mom!" Yelled Sky.

"Yes?" her mom replied.

"How much time do I have left?"

"You have 1 hour," she replied.

Sky looked at her notepad, then looked at the clock.

Excitement crossed Sky's face. Now she knew what time she was leaving.

Sky ran upstairs. There was her favorite outfit, with the perfect bows to match.

As soon as she finished getting dressed, she glanced at the clock. She knew it when it had been exactly one hour.

She grabbed Jaylen's birthday gift and ran down the stairs.

Before Sky's mom could yell that it was time to go, Sky was waiting for her. Then they were off to the party.

The

End

"What did you learn"?

_____ seconds = ____1_____ Minute

____60____ minutes = _____ hour(s)

_____ minutes = _____2_____ hours

www.ingramcontent.com/pod-product-compliance
Lightning Source LLC
Chambersburg PA
CBHW051832210526
45473CB00005B/1846